Though the words
in this little treatise
may modify your
view of yourself
and the world,

they do not constitute
a philosophy per se
(though some
might say it opens
the door to one);
nor do they in any way
constitute a religion,
or an ideology,
or a prediction
or prophecy;

they are, basically,
a simple, objective
statement of fact
that you will
acknowledge, or
deny, according
to how the *you*
that you are
came to be.

The Tiny Truth
That Governs
Everything

ISBN 978-1492960911

fabianmelmelgar@gmail.com

Also by
Fabian (Mel) MelGar:

The One Truth
That Governs
Man, The Gods,
The Heavens
And The Universe
© 2006
(Early version of
The Tiny truth
That Governs
Everything)

**Paint The Beast
Pretty** (a cerebral
mystery novel)

FabianMelMelGar

THE
TINY
TRUTH

THAT
GOVERNS

EVERY
THING

The *ifs*, *ands* and *buts* about it:

If the tiny truth of this treatise were not true, you could sing no song, you could smell no rose; you could marvel not at the rise of the sun or the sweep of the moon; if it were not true, no wheel could turn, no gear could grind, no heart could quiver, no mind could wonder; if it were not true, there would be no me and there could be no thee; if it were not true, there could be no clouds and there could be no rain, there could be no streams and there could be no seas; and if it were not true there could be no butterflies afluttering or elephants atrumpeting; nor could there be a little blue orb warming up close to a large ball of roiling hot embers in what is an immense, mostly empty and cold, universe; but this tiny truth, which makes that little blue planet and our place on it possible, is not revered, is hardly even known, though knowledge of it might change our world, might make it better, might make us more open and caring, less dogmatic and opinionated, more rational and accepting; it could nudge our world toward a better place, a more just and equitable world, a world where everyone would have the opportunity to achieve and contribute to the best of their abilities, a world where everyone mattered.

So, here, if you don't already know it,
and with the object of making it
more known, and better understood,
the simple *ins* and *outs* of the
tiny truth that governs everything:

For starters;

there are many theories about man,

and many beliefs about the gods;

many tenets about the heavens,

and many facts about the universe;

but there is only one truth

that governs man *and* the gods, *and*

the heavens, *and* the universe;

and it stands on two simple principles:

one is that all moments in time,

whether eventful or not.

whether regular or random,

whether understood or not,

are caused;

the other, a bit less simple, is that

all motion proceeds in the direction

where the cause precedes the effect,

regardless of the direction of time;

and these two principles

are embodied in this one tiny truth:

Everything as it is
at this moment

was caused by
everything as it was
a moment ago.

That may seem an innocuous little truth;

but consider that everything

as it was at *that* moment was caused by

everything as it was the moment before *that,*

and everything as it was at *that* moment

was caused by everything as it was the

moment before *that,* and the moment

before *that,* and the moment before *that;*

and the moment before *that;*

which means:

that everything as it is at this moment,

was caused by everything as it was at any,

and all, of the unknowable, uncountable,

trillions, upon trillions, upon trillions, upon

trillions of moments that have already been

—the fact, for example, that you are reading

this at this moment was causally determined

by everything as it was twenty six trillion,

seventy five billion, six million, two hundred

twenty eight thousand, one hundred sixty one

moments ago, and also by everything as it was

just thirty three moments ago, and one

moment ago, and all the moments between.

It cannot be otherwise.

And
everything
means

EVERY
THING

—past,
present
and
future—
including, but not limited to . . .

the 7,000,000,000+
humans on Earth,

THE **2,000,000,000,000,000,000,00+**

SUNS IN OUR UNIVERSE,

every
bounce
of a ball,

every
thought
in your
head.

Nothing is excluded,
no moment of time,
no random event,
no man, no god,
no grain of sand,
no living thing,
no anything;

And just as **everything** means **everything**,
caused means **caused**, and
determined means **determined**;

which means —unequivocally—
that **everything** as it is at this moment
was **caused** by **everything**, as it was
determined it would be, the moment before.

That is what it all means; but what it *doesn't* mean, is
that we can look ahead to a specific, future moment
in time and space, and unequivocally predict everything as it will be at that given moment in that given
space; and, though there will be times when we can
predict all, or some, of it, there will also be times
when we can't predict any of it. Basically, whether or
not we can predict, exactly, the state of a specific moment in the future, depends on the complexity of that
specific moment and of all the moments leading to
that specific moment, and how much control, if any,
we have of those moments, and on how many moments there will be between the moment we make
the prediction and that specific moment. When we
look at ourselves in a mirror, for example, we can
safely predict with unequivocal certainty —because it's
a simply deduced fact, without an alternative— that
the person we see there will, one day, die; but at what
exact moment in history that person will die, would

be, because of the complexity of life's moments, and the large number of possible moments till that exact moment, impossible for us to determine; yet, determined it would be, because it will have been caused by all the regular and random moments that preceeded that final moment.

Some skeptical readers may want to claim that **caused** does not always mean **caused**; that the words, *chance* and *random*, may be indicators of events that are not caused; and, if so, our tiny truth would not be universally true. But, contrary to what those readers may choose to believe, the words, *chance* and *random,* indicate actions without **expected***,* or **regularly occurring**, causes, not actions without a cause. Consider, for instance, this hypothetical *chance* event: someone tells you how they bumped into a friend in a foreign city clear across the globe. They think the odds of that being caused simply by chance have to be so implausibly high that its happening has to be thought of as a miracle magically created by their god. Of course, if that's true, that god would be the cause. In any case, the odds of its happening is not at all in the realm of miraculous, no matter whether the word is used in a theistic or atheistic sense. If you search your memories you will recall large and small incidents like that in your own life. I, myself, have had a few of these so-called miraculous incidents.

Once, many years ago, for instance, when I was in the business of demonstrating to the public the attributes of my clients' products or services, I called in a contractor (whose name I plucked out of the yellow pages) to come to my offices to break through a wall and install a door. He came early on a Friday morning and with a crew of two helpers installed the door in a matter of a few hours. When he was done, I paid him and thanked him. Late that afternoon, my family and I flew to Florida to enjoy a vacation at Disney World.

The next morning, as we were working our way through the overflowing crowds on The Magic Kingdom's Main Street, a man suddenly stepped in front of me, and with a look of surprise and disbelief on his face, said, "Mr. MelGar! remember me?" Well, I guess I don't have to tell you he was the contractor that I had met in New York City for the first time the day before; and now here he was in Disney World with *his* family. After a handshake, and the exchange of a few words, he rushed back to them, got lost among the crowds, and I never saw him again. A mathematician could figure out the odds of something like that happening to all of us sometime in our lives; but whether caused miraculously by a god, or by simple chance, it would have been caused by everything as it was in all the moments before the event.

As to, **random**, most of us would think of a random event as somewhat similar to a chance event, in that random events can also be unexpected and unpredictable; but they are different in that they occur much more often, though we seldom notice them unless they are of a catastrophic nature. Also, the complexity of some random events, and their causes, occasionally appear to be inexplicable, and that may cause skeptics to deny —if they haven't already done so— that those random events are determined; and therefore, deny that they are caused by everything as it was the moment before. Those same skeptics may take it further and claim that with all the chaos that exists in the universe, including all the moment to moment interactions of all its tiny atoms and particles and its massive planets and stars and moons and asteroids —with their inevitable, random huggings and tuggings and mergings, and clashings, and their exchanges of matter, each with the other— how can we claim that everything as it is at this moment, which we are usually incapable of defining, was caused by everything as it was a moment ago, the state of which we also cannot define with much certainty. But the searching for those definitions are precisely what science is about; and science could not have proceeded in its search for knowledge of the inner workings of the world without first having discovered that everything in the world, as it is at this moment, was causally determined by everything as it was a moment ago, and that it will cause everything to be as it will be a moment from now.

Before we get entirely away from the subject of random causes, random happenings are, by their very nature, the ones most difficult, and sometimes seemingly impossible, to get to the root of (I'll be discussing one or two of those as we go along), their very nature makes them, to some skeptical readers, the weak links in the argument for causality; but in reality they are the strongest links in the argument; regularly occurring moments do very little in the way of causing other than other regularly occurring moments, ad infinitum; but it is the very nature of random events that they often cause moments that are very different than their preceding moment or moments; and they are the cause of the evolutionary genetic changes that occur in all living things on our planet; and those living things includes us, the homo sapiens of today —the ones who search for the causes of things— and it includes all of our ancestors, including those most ancient, those who scooted from water to land, and back to water; and from sunshine to shade, and back to sunshine again, in their regularly and randomly occurring struggles for survival, till a random genetic event made it possible for one of them to survive on land alone, where began a long chain of occasionally occurring, random genetic events that led, over a span of uncountable trillions upon trillions of moments, to us as we are today, the most advanced, though far from perfect, thinking machine on Earth.

Spontaneous and *accidental* are two other words that some may think indicate events that are not caused.

But the words, *spontaneous* and *accidental*, simply indicate something that happened that *wasn't planned*; they don't mean something that wasn't caused.

And, lastly, to round off this group of words that might be considered as disparaging of the tiny truth, let's consider the words, *luck*, and the hypothetical, *if*.
The word, *luck*, according to Webster's Dictionary, means "The circumstances that work for or against an individual." The circumstances (in the context of this treatise) being everything as it is at a given moment; all of which would have been caused by everything as it was in the trillions of trillions of trillions of preceding moments. As to the hypothetical *if*, there is no room in this treatise for an *if* that says, "*If* only I had done this instead of that, things would have worked out better." Well, they might have worked out better; we can say so hypothetically; but, in reality it couldn't happen, there could be no *if*, we could not have chosen differently, we were what we were and the circumstances were what they were; and were we to recreate that event, to the finest detail —including our ignorance of prior outcomes— a million times or more, we would make the same choice every time.

(When I say everything as it is at this moment was caused by everything as it was many trillions of moments ago, I don't mean to say that it was planned by some ethereal being, or beings, as some may believe —this is not a treatise built on beliefs; it is a treatise based on scientific facts and logic that constitute a truth— I'm simply saying that everything as it is at this moment was causally determined by everything as it was a moment ago; and that, if there is a god, anything attributable to it, at any given moment, was caused by everything as it was in the moments that preceded it. For example, that god's decision to create us, if indeed it did so, would have been caused by the state of its mind the moment before it did so —that god could not have created us and then decide to do so— or that god could have created us by accident, which would have been caused by everything leading to that accident.)

Here are a few thoughts that may help in grasping what this tiny truth might look like in our daily lives:

Your choice of a mate was irrevocably determined before mating even existed. You were what you were, and your mate was who they were, and neither of you could have chosen differently —nor could anyone have predicted it —a confluence of uncountable trillions of trillions of events (many of them random) that began way back in the mist of time, brought you together; and who each of you were, at the moment you met, sealed the union.

If you say a prayer for the life of a loved one who is ill, and that person survives their illness, it doesn't matter whether you think that your god saved your loved one, or you think the doctors, nurses or medicines saved your loved one, or that your loved one's immune system did it, the cure was caused by everything as it was long before that person was ill.

On a clear night, look up at the moon as it circles our planet on its delineated orbit. If you were told that its exact position, at that precise moment, was caused by everything as it was at the beginning of our universe, you could maybe accept that possibility. But if you were to look up on a sunny day and see a little yellow butterfly fluttering by erratically, and were told: that butterfly and its exact position, at that precise moment, was caused by everything as it was somewhere back in the haze of time, you would be loath to acknowledge it, you would think it mad to think it so. Yet it's true, and there is no plausible way it cannot be true.

Having said all that, here are a few disclaimers of sorts: First, the tiny truth discussed in this treatise is not a new insight. Secondly, it has some naysayers. And lastly, contrary to my assertion that everything is caused by everything as it was the moment before, there may have been a moment when it wasn't.

For thousands of years virtually all scientists and philosophers, whether professional or amateur, have known the essentials of the tiny truth of this treatise. Mostly they referred and refer to it as determinism, or causality, or causal determinism. The truth described in this treatise, however, is a simpler, but more encompassing, form of those: an absolute determinism that includes everything from the tiniest cell in your body, to the largest acts of creation and destruction laid claim to by the gods, a form that most pre-enlightenment, and some post-enlightenment, scientists and philosophers dared not, and perhaps dare not, contemplate. There are actually some philosophers and scientists living today who dismiss any form of determinism —especially those forms that claim to be absolute— and they profess to do so strictly on rational, reasoned, logical, scientific grounds.

There are a number of philosophers who appear to oppose determinism on pragmatic grounds. They evidently feel that those of us who are less enlightened than they, would not be capable of assimilating this tiny truth in a constructive way; that they feel determinism should not be thought true because they believe it would be detrimental to our laws and morals to think it so; and they write intricately reasoned volumes asserting some scientific, or logically deduced fault, in it. (I will return to this subjective argument later, at a more appropriate place in the narrative).

The more objective, more technically complex and difficult to grasp, negative arguments advanced by some scientists, and writers of popular science, are based on chaos theory, on Einstein's theories of relativity, and on quantum physics, and are impossible to present and respond to equitably in this simple little treatise. The arguments stemming from chaos theory, and Einstein's theories of relativity, can be readily researched and dismissed (as they relate to this treatise) by anyone who is willing to devote a bit of time in exploring those subjects. However, the arguments stemming from quantum physics —with its Copenhagen Interpretations, its Werner Heisenberg uncertainty principle, its EPR arguments, its John Bell theorem, and numerous other perceptions (again, as they relate to this treatise)— would be, for most non-scientific readers, much too difficult to assimilate judiciously; so I will dismiss those arguments on the simple grounds that though quantum physics is held to be very accurate in its description of the submicroscopic world of atoms and subatomic particles, some actions of those atoms and particles can only be pinned down on the basis of probability and not on the basis of the certainty that some might claim our tiny truth requires; but when we put together huge numbers of atoms —some fifty million, million, million

in just one grain of sand for instance— the probabili-
ties average out consistently and become the predictable actuality that operates at the macro level of
our everyday world of thought and speech, and peas
and carrots, and knives and forks, and telephones and
televisions; and, of more relevance to this treatise, the
skyscrapers and flying machines made of glass and
metal and rivets and grommets and struts and beams;
all engineered using, not the unpredictable quantum
physics of the micro world, but the predictable, classic
physics of the macro world. If this were not so, not
we, nor any scientist or philosopher, would dare to go
up in any of those skyscrapers or flying machines.

Lastly, I have to admit to the possibility that not everything is caused, that there may have been a moment
in time, in the history of everything, that had no cause
because there was no need for a cause, that the moment of the beginning of everything had no discern-
able cause because it is quite possible that there never
was a beginning, that the matter of the universe always existed. I think we can accept, with reasonable
scientific certainty, that the sudden, explosive expansion of existing matter—an expansion known as The
Big Bang—was the cause of the birth of our known
universe; but where that existing matter came from,
and what caused it to be, is not known, and is an open
question. Some religions—the ones that accept that

there was a Big Bang— say it was their god that created the matter that expanded and became our universe. But that brings up the follow-up question of
what caused that god, itself, to be? Perhaps they will
argue that their god, or gods, always existed, and
therefore there was no need for a cause; or they may
argue that their god, who didn't exist, created itself
and the universe from nothing. Well, generally speaking, each of those claims may appeal to some segment
of our population; but, rationally speaking, it would be
less of a stretch, and more in keeping with the simple
scientific nature of this treatise, to just state that the
basic, fundamental nature of matter, has always simply
been to be, to simply exist rather than to not exist;
because if matter's nature was to not be, to not exist,
it would have required the complication of a supernatural force for us to exist and be here to consider
it. So, with that caveat to the use of the word everything —and in the interest of keeping things simple,
rather than getting into what might become the ne-
cessity of exploring, without end, and without hope of
discerning it, the abstruse role of enigmatic antimat-
ter— I'll just keep stating that everything is caused by
everything as it was the moment before, and, mean by
it, only the moments whose existence we can reason-
ably account for scientifically, meaning all the moments
since The Big Bang.

All this about the beginning of everything, and about quantum theory, brings up the question of **predictability**: Nowhere in this treatise is it claimed that because everything in our everyday macro world, at this moment, was caused by everything as it was in all the trillions of trillions of moments that preceded it, everything is predictable. That is not to say, as I've stated before, that short term, simple things are not predictable. For instance: you are standing in the middle of a desert; you have a cannon ball in your hand and you let it go; we can safely predict that the ball will fall to the ground, we all know the law of physics called gravity, we know it because we experience it every time we fall. And while there are many other things that are simple and predictable, there are many more that, though causally determined, are, like the following examples, much too complex to predict:

Get a large barrel and fill it to the brim with dice; take the barrel to the top of a high hill; dump all the dice down the hill's stone face; watch them roll down, tumbling, bouncing off each other and the hard rough surface of the hill; watch as they come to a halt at the flat ground at the bottom; look at the numbers that are face up; they would all have been determined long before dice were thought of, but no gambler or mathematician would dare to attempt to predict them. Nor could anyone replicate, exactly, that roll of the dice.

Think of a maple tree in the autumn, its leaves all or-
anges and yellows and reds. Exactly when each of its
individual leaves will fall to the ground —the random
fall of each leaf dependent on a multitude of factors,
including, but not limited to, the actions of moon and
sun and rain and wind, and the whims of squirrels and
birds and bugs and bears— would be impossible to
predict. Ask any arborist, or physicist, if they could predict even one. Yet the which and where and when of
their fall was determined long before the tree
sprouted from the ground.

A similar example to those leaves can be seen in
the world of atoms and particles discussed earlier:
Predicting when a given atom in a chunk of radioac-
tive uranium will decay (by the emission of an alpha
particle) is impossible to do; but, again, as with the earlier quantum discussion, how many atoms will decay in
that chunk, in a given macro time frame, is predictable.

The results of the preceding, complex examples,
would be, if tried, impossible to replicate on later attempts; not just because of the extreme difficulty in
physically reproducing everything exactly as it had
been at the moments of the original attempts, but because of the impossibility of attempting them at the
same moments in time as the originals; every specific
moment, in its universal entirety, lives just once,

it causes the next moment, and then it is gone,
it is history, and cannot be physically resurrected.

And while you're in the act of mulling that over in
your mind, think about the brain where that thinking is
going on —a brain that contains some one hundred
billion neuron cells in which are stored the totality of
your mind and its knowledge— how did those billions
of individual neurons, with their connecting synapses,
learn and evolve and adapt, from the moment you
were born, as they were exposed to the world they
were born into? Could anyone who knew you at birth
have predicted exactly when your mind would figure
out that one plus one always equals two and not the
three that some may have indoctrinated you to believe? At what moment you figured it out, if in fact you
ever did, would have been determined by everything
as it was long before man figured out they could
stand on two legs.

So, now that I've acquainted you, as best I can
in this little treatise, with the tiny truth that governs
everything; how, you may ask, would that knowledge
change how we view ourselves and the world?

Well, to put it bluntly, and to get quickly to the crux
of the matter, our little truth means **we are not
self-made,** it means we did not create the we that we

are at this moment; it means we were created by everything as it was in all the moments that preceded this moment, all the way back those trillions of trillions of trillions of moments ago; it means that we were created by a god and/or every one of the very long list of our direct ancestors, and by all that this "we" that is us has been taught, and experienced and learned, from birth till now; and the exact person we are at this moment —to the tiniest detail— **is the only person we can be.** And that means **we do not have free will** —that is, we are not capable of making choices other than the ones we make— it means we have will, and it's ours, but **it is not free to choose other than what it chooses.** It may feel to us, as we live our lives, as if we *are* making choices; it may feel to us as if we are choosing between two or more options. And we are, but the actual choice we make at any given moment **is the only one we can make.**

We humans, masters of our world, don't want to know that we can't make free choices, that the next choice we make will not be of our own making, not made by the free-thinking "me" that we think we are, that it will have been made because of a long chain of past events beginning way, way, back, and leading to the moment we make the choice. We feel that to admit that —**that we have no free will**— means we would be admitting that we are nothing more than mindless automatons; that we'll obey who we're told to obey; that we'll believe everything we're told; that we'll do anything we're told; that we'll love who we're told and hate who we're told; that we'll believe the most childish fantasies; that if we have no free will, no will of our own making, we will be nothing more than placid, malleable sheep induced to follow any Pied Piper who promises to lead us to green grass and cool water. And we would be right, all those things can, and have, happened; **but they don't happen because we don't have free will; they happen because we think we do**. That is how the Pied Pipers of the world get their followers: by convincing them that they follow of their free will.

Most of us believe we are free to think whatever is thinkable, and choose whatever is chooseable; and we believe it because we think there is a separate, inviolate, autonomous "me" that resides in us, a distinct "me" that's there from the moment we're conceived, a "me" that is virginal and pure and not influenced by its creator—be its creator a group of genes or a god or both—an individual "me" that's able, from the moment we're born (or perhaps before), to use its supposed free will to make choices, to discern whether incoming information is factual truth or spurious fiction or something between. It is, however, difficult to understand on what basis we could make such free choices since our virginal minds would be virtually empty of any sort of input or knowledge on which we could base our choices; and if it is not virginal, it must have been impregnated by our god or/and our genes, with some at least rudimentary knowledge; and there-fore the choices we make would be based on what we have been impregnated with and are not ours to will freely, but are dictated by our god and/or our an-cestral genes. But, there are others of us who believe that except for our genes —which determine our physique, the construct of our brain, and, possibly, some, or many, of the rudiments of our mind— there is no autonomous "me" when we are conceived, only the gradual emergence of a conscious awareness of ourselves as we develop, first in our mother's womb, and then out in our mother's arms, and, finally, the en-compassing arms of mother Earth. And, lastly, there are those of us who believe that we are born as naked in mind as we are naked in body, with no sense of anything at all till we learn from the world we are born into.

But it doesn't matter whether we think of our "me" as our "soul" or just our "me," or whether we believe we are born with a blank brain, or with a brain already having begun to process and use incoming informa-tion to form a mind, or with some amalgam of those beliefs; when we are born we need someone with experience of the world to be our guide, to inject our spongelike brain with contemporaneous knowledge that will teach us how to learn and survive in the specific world we are born into. And that someone is our parents and our older siblings and anyone else who is active in our lives; and it is they who teach us about that world, and they teach us according to their beliefs and understandings; and since, initially, there is no one in our lives, as powerful as they, who can contradict what they are teaching us; their ideas —remodeled by our needs and innate abilities, if in fact we are born with any—become imprinted indelibly in our minds. Then, as we grow and we come in contact with the wider world, our imprinting is strengthened, or modified, by our teachers and our playmates, and, later, also by books and magazines and newspapers, music, movies, television and the internet. And, after all that —when we are finally adults—the "we" that we are is the product of all of that; we are the product of our biology, our experiences, and all that imprinting; and, unless we have been taught to keep our mind open and critically questioning, that imprinting becomes set, like concrete, in our brain. But regardless of whether our minds are set, like concrete, or our minds are still pliable, what we are at a given moment is what we are, and we cannot be otherwise; and what we do at that moment cannot be other than what we do. But when we *think* there is an inviolate 'me' that is not ruled by what we've inherited, and —more apropos here— what we've been imprinted with and experienced, we don't question what we are doing, we don't question the indoctrination we have received as children and the imprinting we've continued to receive throughout our lives; and that may cause us to do things that are not in our best interest, or in the best interest of those we care about, or in the best interest of the society we live in, or that of the world at large. And that is the danger. The danger is not the fact that we don't have free will, but that we think we do; and that because **we think we do**, we don't develop the tools of critical thinking and questioning that would allow us to make informed and examined choices instead of the sheep-like, herd-like, choices that dogmatic indoctrination leads us to make.

And the biggest danger of all is that, because we believe we have free will, we pass that belief onto our children and we don't teach them to bring an open but critically questioning mind to the consideration of new ideas. Unfortunately, that leaves them helplessly stuck, for a lifetime, in a bog of antiquated, narrow beliefs, or captive to a whirl of new fads foisted on their unquestioning minds by promoters of fashionable products or philosophies. And for those of us who think we have free will because a god gave us our soul, and, with it, free will, it must be that when that god speaks of free will, she or he simply means that he or she gave us the ability to make choices; because, if that god knows all, she or he knows that we can make choices, although we may make them subconsciously (as Daniel Wegner, and Sam Harris, and Mark Twain, state in their respective books, *The Illusion of Conscious Will*, *Free Will*, and *What is Man*), but that the only choices we can make, consciously or subconsciously, are the ones we make; and that is not the free will that most of us would like to believe we have. **The "we" that we are at the moment we make the choice, and the event that requires the choice to be made, determines the choice, and that choice is the only one we can make. Our brain, with its mind, or soul, is the only tool we have that we can use to make those choices, and it can only be as it is, it can only be as it was crafted for us.**

(That does not mean, however, that we are all the same, as if we came off an assembly line; it does not mean that we don't have a will that's our own. Quite the contrary; we, all seven billion plus of us, are all gloriously unique: we are all, including identical twins, different at birth, and, except, possibly, for twins, at significantly different moments in our family's and world's history; we may live in different areas of the world, speak different languages, and have different belief systems, and be in different socioeconomic and political environments, and have different experiences; all of which possible combinations, if totaled up, would add up into trillions upon trillions upon trillions of trillions of different possibilities, making it quite impossible for there to be, or to ever have been, or ever will be, two of us humans exactly alike, with identical wills.)

Many of us think we can change on our own the "we" that we are; but, in reality, it is quite difficult.

It is difficult even if our minds are not set like concrete, and even if we want to change the "we" that we are. Only if we've learned the value of having a critically questioning mind, and being open minded —of not being dogmatic— and if we've been exposed to mind-altering knowledge, or experiences, can we change who we are. Sometimes a new bit of information that we are exposed to can fill in the gap between two seemingly unconnected pieces of knowledge that we already possess, creating something new in our minds —a new vision of the world for instance.

But sometimes an experience can dramatically, and forever, change us and our vision of the world regardless of the state of our mind, our will, or our wishes. Take the example of a strong young man who thinks himself heroic: One day he is in a local grocery when a hooded man suddenly points a murderous looking assault weapon at everyone in the store and tells them he will kill anyone who moves. As the desperate hold-up man tells a clerk to get him all the money in the cash register, our hero decides —that is, he chooses, because of who he is at that moment— that at the first opportunity he will jump the hold-up man and wrest away his gun. Suddenly, at the sound of a police siren, the hold-up man turns to look out the window, and our hero is about to jump him when another young man, also envisioning himself a hero, dives at the hold-up man. But before this new hero can get his hands on him, the hold-up man quickly turns and fires his assault weapon at him, effectively blowing his head to pieces and killing a neighboring person in the process. Suddenly, on witnessing this, our original hero becomes a different person than he was. Some might say he has become a coward; but, more likely, we should say he has been profoundly changed, and not by use of his will but because the experience has made him a different person, a pragmatic realist; and when the killer escapes out the back and is never caught, our hero may also become a cynic—one less able to have faith in society's ability to protect him.

Though it may be hard for some to accept, everything about free will applies to the so-called immaterial world as much as it does to the material one.

Those readers who may have the most difficulty

accepting the truth of this absolute determinism, are those who believe in the existence of ineffable, transcendental, ethereal things like goblins and ghosts and spirits and souls, and gods and angels, and the efficacy of prayer, and heaven and hell, and in the ability of some humans —those with extrasensory perception— to see into the future and the past and to contact the souls of the dead. Those readers feel certain that those things have no material aspect, that they are supernatural, and they can't imagine how *the tiny truth* could apply to them. And, as to their god, they will say, "Our God is all powerful, he can do anything, he has free will." That a god is all powerful in every other sense may or may not be true, but our simple little truth is an unequivocal truth that applies to gods as well as to ourselves, and to souls and ghosts and spirits and angels and all other things which, if they exist, appear to have no material aspect. It can not be otherwise; neither gods, nor angels nor any other immaterial being can do something —unless by accident— before they, consciously or unconsciously, think about doing it; they can't scratch their butt and then decide to do so. And the truth is that not even a god can alter that any more than we can. A god cannot have a will that is not predicated on forward sequential thought, because without a forward step-by-step progression of conscious or unconscious thought and action, a god would not be able to accomplish anything

—he, or she, could create no us, no universe, no anything. Many people think of gods and the soul as being supernatural and immaterial, but, if the heavens and the gods and the soul do exist, they are not nothing, they are something. The fact, as claimed, that these things are not analogous to the material world does not mean they are not subject to our tiny truth. And if there is a hell, it also is not nothing, it is something, and it too must observe that truth. Neither in heaven, or hell, or earth, can you pick your nose, before you consciously, or unconsciously, think about it. If this one encompassing truth weren't a fact, the heavens would be incapable of being, and gods' minds would be incapable of thought, and the same would be true of our "soul" or our "me." Even the creators of comic book superheroes, are intelligent enough not to create heroes with that incongruous ability.

At this point, you may be saying to yourself that a lot of this about the one truth may be true, but that, nevertheless, you feel that you have free will and that you can change your mind anytime you want to and do things differently than you might do otherwise. To prove it, one Saturday morning you get up earlier than you normally would, deliberately changing your routine. You take a longer, hotter shower than normal; you get dressed in exercise clothes you haven't worn in years; and you go for a morning walk —something you've never done— in a wooded area that you've never been to. You follow a well-worn path that seems to circle around through the woods and back to where you started; but after you've been walking a while you come to another path —a path, not so well worn, that seems to turn off deeper into the woods— and in order to be different, in order to prove to yourself that you have free will, you take this less traveled path; and sometime later you suddenly, unexpectedly, come upon a big momma bear caring for her two cubs. She turns your way and instantly determines that you are a threat to her cubs and she is on you before you can react, knocking you down, slashing your throat, going for the jugular. Then she leaves you there bleeding to death. And so, you would say, if you were capable of saying anything, "See, I changed what it was determined I would do, and I changed the moment it was determined I would die." But the fact is that you wouldn't have changed anything. The fact that, hypothetically, you had been in a bookstore and were intrigued by an earlier version of this treatise, and had begun to read it, is because of who you might have been as a person at that moment in your life: the kind of person that led you to be in the bookstore in the first place, the kind of person who wanted to read that book, and the kind of open-minded person who liked to disprove other people's assumptions; and, because of that, you got up early that morning and did all your morning rituals differently and went for a walk in a place you had never walked before and took an alternate lightly used path deeper into the unknown and found yourself face to face with death. That's the person you were when you ostensibly picked up that book, and you could not have done anything different; everything as it was at the moment of your death would have been caused by everything as it was the moment before, including every thought and emotion in your brain and the existence of that book, that path and that bear. And, whether this were true, or just the story that it is, the moment of your real or fictional death would have been determined a long time ago, long before you entered that bookstore, long before man was man or bear was bear.

Or, by now, you may be convinced that you can't change who you are or the things you do, and you may say to yourself, "If everything that will happen in my future life is causally determined and unchangeable, I might as well stop striving to accomplish anything and just sit on my hands and do nothing and wait and see how things play out." Well, if that is who you are and who you want to continue to be, that is what it is determined you will do, and you will bear the consequences of that; but if, instead, you go on striving to do something productive with your life, then that is what it was determined, by all the prior moments of your life, that you would do; and the life you will continue to lead will be a result of that. In either case, you can only act according to what you are and what your world is, which was all determined those trillions of trillions of moments ago. The preceding also applies to any god that may exist. That god might one day choose —in order to be different, in order to demonstrate that he, or she, has free will— to terminate our universe and create an entirely new one. If that god does so, it's because of who that god is at the moment that god makes that decision, and that god would be the cause of all the effects of it. That god could, alternately, because of who that god ostensibly is, completely eliminate our universe as if it had never existed, as if it never occupied a slice of time, or a space, in that god's own memory. But if that god did so, that god would then be at the exact same state he or she had been at when she or he first created our universe, and that god would do everything the same as she or he had done it before; creating it exactly as he or she had done it before, and causing you to be reading these exact words just as you did before.

Some final considerations and conclusions:

"So," you may ask, since I haven't specifically brought it up, "If we don't have free will and can't make choices other than the ones we make —choices based on the genes we inherited, and the indoctrinations and influences that we absorbed from the intra-community we grew up in—how are we responsible if we break the law and harm a fellow member of the very community that failed to raise us to be a caring, empathetic adult, an adult that would be an asset, and not a burden, to it? How is it justified that we be punished for that? Shouldn't that community, itself, have been considered guilty of causing that harm, guilty of failing to raise us to be good citizens who would not do a fellow member any harm, other than in the act, if necessary, of defending one's self, or one's community, from a member gone wrong?" Well, that's true, but raising everybody to be good citizens is an impossible task if our community does not subscribe to an egalitarian ethic, if it doesn't treat all its members as equal citizens, if it does not provide each of us an equal opportunity at success in life, if it doesn't encourage, and provide the means, to allow each one of us a chance to enhance our abilities according to our innate attributes and desires, and of using that learned proficiency to contribute to the wellness of our community; and if it doesn't encourage us to dream and soar as high as our wings will take us; and if it doesn't encourage entrepreneurship and the growth of industry and living-wage jobs, and lastly if it doesn't provide for the health, and especially the mental health, of its members. And all that, while encouraging us to take responsibility for ourselves, our families, our friends, and our neighbors, and only turning to the community, as a whole, when the need is beyond the scope of the individual. And all that, while reminding ourselves that we, all seven billion of us, are one community, living together on one little blue planet; and that we are each responsible, individually and conjoined, for keeping it livable, for keeping it beautiful and bountiful.

But, having said that; successfully raising all of Earth's children to be good members of our community simply by creating a more egalitarian culture, is expecting, in the near term, more than man, individually, and as a community, can deliver; and so, we must continue to demonstrate the punishment that those who commit a crime subject themselves to. Unfortunately, how we, as a society, decide to dole out that punishment, and the language we use in the sentencing of it, is directly attributable to our attitude towards the less fortunate among us, and especially to those of the less fortunate who turn to a life of crime as a consequence of having grown up in disadvantaged circumstances. Those of us who are fortunate enough to have a productive and rewarding life, would do well to remind ourselves that though we are all considered to be equal in the eyes of the law, we are not equal in our ability to abide by the law; and that any combination of personal, mental, physical, educational, or environmental deficiencies, may cause any of us to engage in criminal behavior; and rather than demonizing such acts by terming the persons who commit them as vile and monstrous and depraved and diabolical and unredeemable —ugly words that prosecuting attorneys and the media are fond of promoting; words, among others, that are maliciously meant to incite the passions, rather than the intellectual cool–headedness required of jurors and judges— we should consider that the use of less inflammatory language could lead to fairer trials with more just and equitable outcomes.

As to the role of religion in all of this, there are those who claim that exposing the truth that we don't have a will that is free to choose other than what it chooses, would, if believed, destroy a religion's ability to use their promise of heaven, and their threat of hell, to deter, wrongdoing. Whether there is any truth to that claim is impossible to say, since, in general, many, if not most, humans have seen fit not to be concerned about any of their gods' promises and threats; witness all the wars and killings and brutalities and crimes and inhumanity perpetrated across the ages and around the globe; and, this, despite the fact that many, if not most, people who have committed those atrocities —which includes the abomination of slavery— are professed people-of-God. And it is quite impossible to assess how many people-of-God who don't commit those atrocities, deter themselves from doing so out of a fear of their god's judgement. And if they act humanely, there is no way for us to know that they wouldn't be just as humane without fear of their god; witness the many people who are not people-of-God, who are humane. After all, we Homo sapiens have —gods or no gods— empathy for our fellow man; we are capable of donning each others' skin and attempting to understand how each other feels; we are, after all, kith and kin. But none of that matters; when it comes to our opinion of it, we will accept, of it, what our mind—a mind not of our own making—chooses as to whether religions' claims, that we possess free will, are valid; that we are able to make a choice other than the one we make, to choose other than what we choose, to choose between doing good or doing harm. If you have a firm belief in one or the other of the gods who ordain the concept of free will, and you succumb to its bribes of heaven, and its threats of hell (the equivalent of the old carrot-or-stick choice imposed on a horse to get it to do our bidding), then you probably are someone who was indoctrinated early in that belief, and, or, you are someone for whom the idea of spending a lazy eternity in heaven has a powerfully addictive appeal. Unfortunately, there can be no such thing as your being able to choose other than what you choose, no such thing as free will; nor can your god claim to have given it to you, or to itself. Every choice you make is influenced by all the prior things that occurred in your life, including those that occur the moment before you make the choice; but, if you should be fortunate enough, all those influences you have been subjected to, will have been, in the main, good influences; and, if there is an actual heaven, and you have a belief that there is, you may very well find yourself there one fine day; though it won't be because you had a will that was free to choose other than what it chose, but it will be because of your good fortune to have been born within a caring, embracing society that provided those necessary, good influences. Finally, on this question of religion and its advocacy for free will, it is difficult to grasp why some religious founders would distort the truth, and assert that, beginning with Adam and Eve, their god gave its human creations free will, when they must have known that if their god was the all-knowing god they believed it to be, that god would know the impossibility of free will. What seems, more likely, is that their god did not make that impossible assertion; it may be more likely that it was the fault of some errant, ancient scribe who wrote down what he mistakenly thought he had heard his god say; or what he had wished his god to have said. Or perhaps that scribe took it upon himself, to make that claim of free will, as a means of justifying his god's unfair, over-reaching punishment of the innocent, childlike, naive, trusting, and curious Adam and Eve, along with all their progeny; because they, ostensibly the first man and woman, disobeyed him and ate the fruit of a tree with the enticing name of *The Tree of Knowledge*. But, as I've said in a prior context, let's not waste our time wrestling with ourselves over the truth, or falsity, of the Adam and Eve fable; it is not worth the effort; we are all vested in what we believe of it; we have been taught it by those who had influence in our young lives, and by what we subsequently learned, on our own, from our experience of the environment that we navigated in —all modulated by the hormones that are, at every moment expressing themselves within us— and all that, together, will always attend the degree to which we believe that ancient fable; as it will the degree of empathy we feel for our fellow man; and will direct us in the choices, good and bad, that we make at every moment of our lives.

And, speaking of empathy, there are some of us who don't possess any empathy towards our fellow man regardless of our views on heaven and hell, either because we were born with a physically damaged brain incapable of empathy for others; or because we were indoctrinated—by word or by example—with the cynical idea that our fellow humans are not worthy of our concern; or because our minds were traumatically damaged by having experienced, or witnessed, horrific physical or verbal abuse that produced in us an unempathetic view of others—as might the possibility that we grew up in a very harsh, very competitive environment where our very survival depended on winning, at all costs, the daily battle with others. When we see people born deformed in mind or body, we generally accept that their inabilities are not their fault, that they are not responsible for their failings; we feel sorry for them, we feel empathy for them, we try to help them—we may even grant some empathy to those unfortunates, mentioned above, who lack any sense of empathy— but when we see people who grow up handsome and sturdy, and apparently healthy in body and mind, make failures of their lives, we denigrate them for that failure; we accuse them of being spoiled, of being physically lazy and mentally indolent, we say that they could have made a good life for themselves if only they had gotten up off their butts and made the attempt, if only they had stopped hanging around without purpose, and done something useful with their lives. For those people, we have no pity, no empathy—especially if we, ourselves, once had some drawback we overcame— but they too are in need of empathy; for they were created with advantages but without the full complement of mental means or necessary conditions that would have allowed them to live the happy, productive life that was hoped for them. The truth is that we all come to adulthood crippled in one way or another; meaning that there are no perfect humans among us. Not that we can agree on what a perfect human would be like. Would it be female, or would it be male? Would it be pinkish, or would it be brownish, or reddish, or yellowish, or even brushed aluminumish? Would it be tall or would it be broad? Would it be fast or would it be strong? Would it be an adventurous seeker of progress or a content seeker of sameness? Would the perfect human be the one who bows to the sacred or the one that stands up for reason? Would the perfect human be prudent, or would it be otherwise? Would the perfect human be someone who revolts at the idea that they be cousin to the ape—one who may resent even being thought of as an animal? And, would this perfect human's index finger be the longest finger in its hand, or would its middle finger be the longest, or for that matter, the ring finger? Would it devour spinach and broccoli for nutrients, or would donuts satisfy its needs, or would it prefer to devour raw, blood-dripping meat as did its ancient ancestors? In actuality, many of us probably believe we know what the perfect human may be like, we may even somehow come to believe—when a less flattering portrait may, more likely, be true—that the perfect human, or a close relative of it, is the one that stares at us in the mirror. So the best we can do for our imperfect, perfect selves is to strive mightily, as best we can, to tamp, or elevate, our egos, as needed,

and do the best we can to make ourselves and our
world better, but to accept that none of us are
the perfect human, that we all have limitations;
and so, emulate a well known icon of yore,
who, acknowledging his attributes and
his limitations, was wont to proudly say,
"I yam what I yam, I'm Popeye, the sailor man;"
for we too are what we are, and the world is
what it is; and all we can do is to embrace
that truth, so that we can know, when
we are dying —as we all must— that whether
we are lying in a prison cell, a rented room,
or a mansion in the clouds, we did the
best we could with what we were given
by the life and the world we inherited.

...No *ifs, ands,* or *buts,* about it.